METEORITE!

The Last Days of the Dinosaurs

Richard Norris

RSVP®

RAINTREE
STECK-VAUGHN
P U B L I S H E R S
A Steck-Vaughn Company

Austin, Texas

www.steck-vaughn.com

To the K/T meteorite—without it,
life just wouldn't be the same!

Steck-Vaughn Company

First published 2000 by Raintree Steck-Vaughn Publishers,
an imprint of Steck-Vaughn Company.

Library of Congress Cataloging-in-Publication Data

Norris, Richard. 1959–
    Meteorite! : the last days of the dinosaurs / Richard Norris.
    p. cm.—(Turnstone ocean explorer book)
    Includes bibliographical references and index.
    ISBN 0-7398-1240-8 (hardcover) 0-7398-1241-6 (softcover)
      1. Catastrophes (Geology) 2. Extinction (Biology) 3. Dinosaurs.
    I. Title. II. Series: Turnstone ocean explorer book.
    QE506.N67 2000   99-27285
    576.8'4—dc21   CIP

For information about this and other Turnstone reference books and
educational materials, visit Turnstone Publishing Group on the World
Wide Web at http://www.turnstonepub.com.

Photo credits listed on page 64 constitute part of this copyright page.

Printed and bound in the United States of America.

1 2 3 4 5 6 7 8 9 0 LB 04 03 02 01 00 99

CONTENTS

CRASH LANDING

"First came three explosions that rattled windows and startled cattle. Then a black, mushroom-shaped cloud of smoke spread across the sky."—Richard Norris

"When I sit down to think about it, it was kind of scary," said Brodie Spaulding. He was outside his home talking to his friend, Brian Kinzie, when they heard a low-pitched whistle or sort of hiss. Next, they heard a thump as something landed about 4 meters (about 12 feet) away. When they went to see what it was, they found a warm, fist-sized black rock, sunken into a small hole in the grass. It didn't look like any rock either boy had ever seen before. Shaped like a kind of pyramid, it looked like something that had split off a larger stone. But the oddest thing of all was the rock's color—jet black and polished, as if it had been painted or maybe cooked in a furnace.

The rock was a meteorite. It was probably one piece of many formed when a rock from space splintered apart in the sky over Noblesville, Indiana, about seven o'clock in the evening of August 31, 1991.

Rocks from space enter our atmosphere every minute. Many of them have broken off from asteroids, which are like tiny planets made of rock or metal. Other space rocks have split off from comets made of ice and rock.

(above)
The meteorite that Brodie Spaulding (right) and Brian Kinzie found wasn't large (inset), but it was rare. No other pieces of the Noblesville meteorite were ever found.

(left)
Meteorite showers light our skies from time to time as rocks from space burn up in the earth's atmosphere.

5

Coming at You!

Meteoroids, meteors, and meteorites all describe cosmic matter. Meteoroids are small pieces of cosmic matter that move in space. Meteoroids travel at amazing speeds—sometimes more than 2,500 times faster than a baseball pitcher's fastball. A meteoroid can travel around the world in 3.8 minutes!

Once a meteoroid enters the earth's atmosphere, it begins to heat up and burn. As dust evaporates from the burning meteoroid, a streak of light is produced. This streak of light is called a meteor, or a shooting star. It looks like a glowing tail behind the meteoroid's head. Meteors come in a range of colors, from orange-yellow to blue-green to violet or red. The color depends on a meteoroid's composition and temperature.

If a meteoroid reaches the ground, it is called a meteorite.

When one of these rocks, called a meteoroid, enters Earth's atmosphere, it begins to heat up. Nearly all meteoroids burn up as they travel through our atmosphere. Some produce sparks and flames bright enough to be seen during the day. Sometimes a meteoroid makes it all the way to Earth. When it hits the ground, like the one found by Brodie and Brian, it is called a meteorite.

Although meteorites strike Earth all the time, they hardly ever hurt anyone—but Edith Hodges was an exception. She was taking an after-lunch nap on the couch in her living room one day in 1959 when she awoke to an explosion. Edith threw off the two thick quilts covering her and saw that her side and arm had been badly bruised. On the floor by the couch was a large rock. The mysterious rock weighed almost 4 kilograms (8.5 pounds). Like the stone found by Brodie and Brian, it was shiny and black, with flat edges. The rock had shot like a cannonball through two sheets of 2-centimeter (¾-inch) plywood in the roof and ceiling, bounced off a radio, and sailed across the living room to hit Edith. By a strange coincidence, the Hodges's house

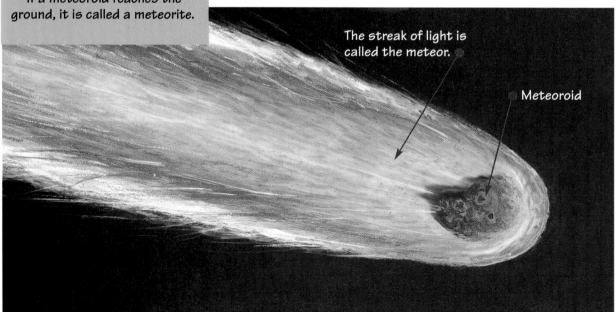

The streak of light is called the meteor.

Meteoroid

This woman's car was struck by a piece of a meteorite called the Peekskill Fireball. The fireball streaked above Peekskill, New York, on October 9, 1992.

in Sylacauga, Alabama, was located across the road from the Comet Drive-in Theater!

The approach of the meteorite was noticed that day by many people around Sylacauga. First came three explosions that rattled windows and startled cattle. Then a black, mushroom-shaped cloud of smoke spread across the sky. Many people thought an airplane had crashed.

A meteorite does not have to actually hit the earth to cause destruction. In the early afternoon of June 30, 1908, a meteor exploded about eight kilometers (nearly five miles) above the ground. Woodcutters and goat herders in eastern Siberia, a place in Russia, saw a bright flash in the sky around one o'clock in the afternoon. The flash was followed by a powerful earthquake.

One witness reported, "The sky split apart and a great fire appeared. It became so hot that one couldn't stand it. There was a deafening explosion, [and my friend] S. Semenov was blown over the ground across a distance of three sazhens [about six meters or about twenty feet]. As the hot wind passed by, the ground and

After a comet exploded over Siberia, trees lay flat and bare, their tops pointing away from the center of the explosion. The damage covered an area larger than half the size of Rhode Island. But the explosion did not leave a crater.

To retrieve samples to study, researchers go on meteorite hunts. One of the best places to hunt is Antarctica. It's easy to see meteorites in the flat whiteness of Antarctica, and the ice helps preserve rocks from space. Here, Guy Consolmagno of the Vatican Observatory examines a meteorite before it's collected.

the huts trembled. Sod was shaken loose from our ceilings, and glass was splintered out of the window frames."

The blast must have been tremendous. Researchers now think the explosion was caused by a comet, or "dusty iceball," burning up above the forest. They estimate that the object weighed about 100,000 metric tons (about 110,000 tons) on entry into Earth's atmosphere, or about as much as a fully loaded freight train. They also estimate that it had the explosive power of about 40 metric megatons (about 44 megatons) of TNT, or around 2,000 times the power of the Hiroshima nuclear bomb.

Though most meteorites that land on Earth cause little or no damage, it's not hard to imagine that a really big meteorite landing in the wrong place could be very destructive. If a meteorite one kilometer wide (six-tenths of a mile) landed on Earth, it would not only destroy everything for miles around, but it would also blast dust and broken bits of space rock high into the sky, perhaps even into space. This dust and debris would have an even greater impact than the meteorite itself. The dust and

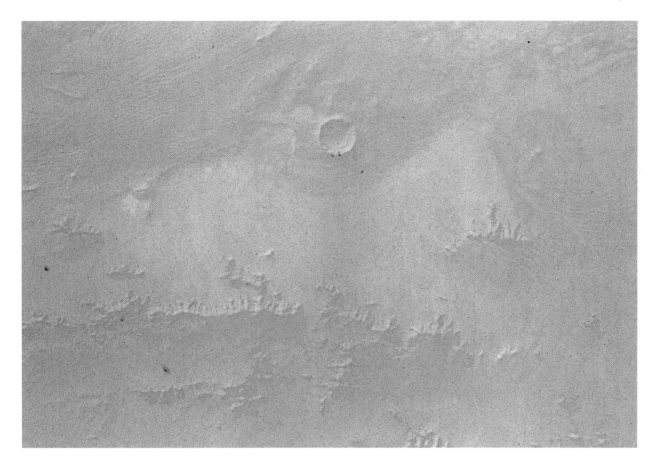

fragments could keep sunlight from reaching Earth's sur-face, drastically changing the planet's climate. The fail-ure of crops, and the food shortages that would follow, could lead to widespread starvation. An even bigger meteorite—several kilometers in diameter—could be so damaging that it could wipe out many forms of life.

Fortunately such events are rare in Earth's history—but more and more evidence suggests that just such a dis-aster ended the reign of the dinosaurs and made way for our own human beginnings some 65 million years ago.

Craters formed by meteorites are easy to spot in flat, bare landscapes like the desert, especially when viewed from great distances. In this shot taken by a space shuttle, the almost perfectly round rim of an impact crater in the southern Namib Desert in southwest Africa is clearly visible.

THE DINOSAURS ARE GONE

"We had clues, but we were missing a key piece of evidence—the certain knowledge that a meteorite impact and a mass extinction happened at the same time."—Richard Norris

Perhaps no group of creatures has such widespread appeal as the dinosaurs. These amazing beasts ranged from the gigantic monsters of science fiction movies to quick, bird-like creatures that caught dragonflies on the forest floor a hundred million years ago.

Dinosaurs lived on Earth for some 165 million years. During that time they developed tremendously varied ways of life and managed to spread over nearly the entire planet. Although some of the dinosaurs' descendants, including birds, are still with us, the dinosaurs themselves are gone. What happened?

We know that the last species of dinosaur lived in what is called the Cretaceous period, which extended from about 144 million to 65 million years ago. Dinosaurs were extinct, or had died out, by the beginning of the next period, called the Paleogene period. The time between the Cretaceous and Paleogene periods is usually given the nickname "K/T Boundary" ("K" stands for "Cretaceous" and "T" stands for "Tertiary," which is an older name for the Paleogene period). Researchers think it was during this time that the dinosaurs vanished, along with many other forms of life, including much of the life in the oceans.

Dinosaur fossils like the skeleton of a *Tyrannosaurus rex* at left are very rare. Only about two dozen complete *Tyrannosaurus rex* skeletons have been discovered. Even dinosaur tracks like these above can be the find of a lifetime for a scientist.

Geologic Time

Geologic time is divided into sections called periods. The period that includes humans is only about two million years long. In geologic time, that's an instant—Earth's history covers more than four billion years.

You can actually see periods as layers in the earth. Over thousands of years, sediments, such as mud and sand, pile up in layers. These layers also hold fossils of the plants and animals that lived at the time the layers were made. By studying the material in a layer, scientists can learn more about different life forms and conditions on Earth at the time the layer formed.

The boundaries, or lines that divide periods, show times of change. Some boundaries match times of mass extinctions. Perhaps the biggest mass extinction of all is at the boundary between the Permian and Triassic periods, when possibly as much as 96 percent of all species vanished.

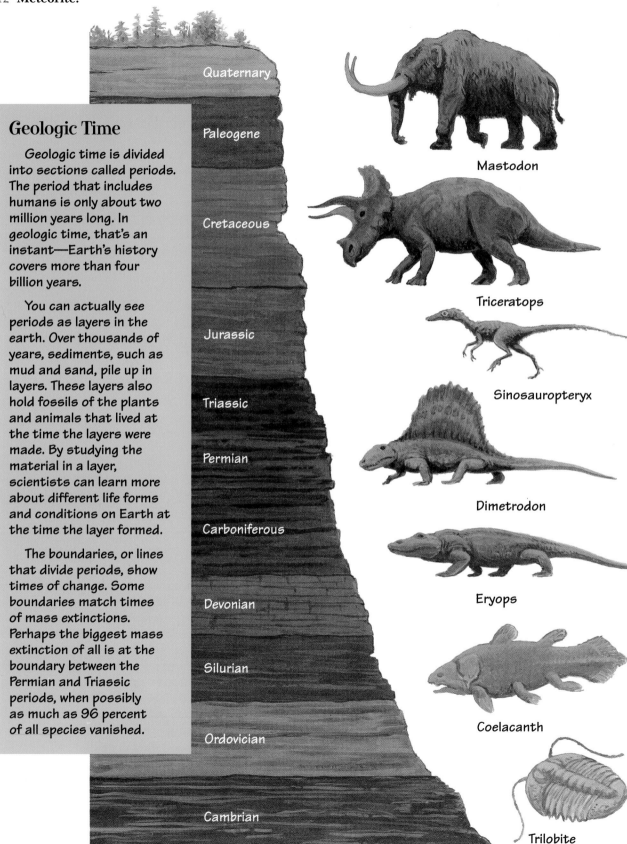

Quaternary

Paleogene

Cretaceous

Jurassic

Triassic

Permian

Carboniferous

Devonian

Silurian

Ordovician

Cambrian

Mastodon

Triceratops

Sinosauropteryx

Dimetrodon

Eryops

Coelacanth

Trilobite

Between fifty and seventy percent of all species in the late Cretaceous period became extinct by the beginning of the Paleogene period. So many species disappeared that this event became known as a "mass extinction," the end of existence for a large number of plants and animals. There have been several mass extinctions in the history of our planet, and one of the most destructive extinctions in the past half-billion years was the K/T mass extinction. Why it happened and how such widespread and numerous creatures as dinosaurs could have disappeared have remained unsolved mysteries of science.

The hillside at Dinosaur National Monument in Utah was so full of dinosaur fossils that it was covered and made into a museum.

There have been all kinds of explanations for the disappearance of dinosaurs, as different as the theory that mice ate dinosaur eggs or that an exploding star showered Earth with radiation, making it impossible for large animals to reproduce. But as scientists began to discover that many other kinds of life disappeared at about the same time as dinosaurs, it became clear that there had to be another answer.

In the 1970s scientists learned that the extinction of a large number of ocean species occurred unusually fast, happening in a few hundred thousand years or even less. That may not sound very fast, but it is. A hundred thousand years is only a tiny part of the history of our planet, even if it's a long time for us. What could kill half of all species on Earth that quickly? Most scientists thought that Earth's climate must have changed faster than animals and plants could adjust. One popular theory was that the climate change was caused by giant volcanic eruptions and changes in the level of the oceans. This theory is still accepted by some scientists, but major volcanic eruptions and sea-level changes have often happened in Earth's history and are rarely connected to mass extinctions.

In 1980 Louis Alvarez of the University of California

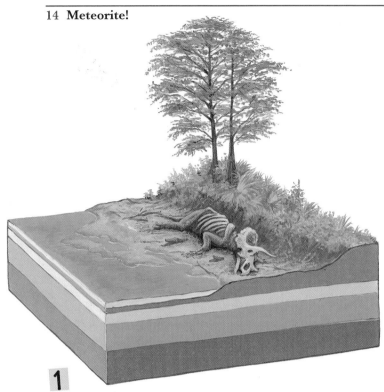

and his son Walter, along with other researchers, proposed that a meteorite impact caused the K/T mass extinction. If they were right, the extinction would have happened in only a few weeks or months, and many life forms would have disappeared in a matter of days. Many scientists ridiculed the theory. They argued that the extinctions took too long to have been caused by a single meteorite. Fossil records we have so far suggest that dinosaurs vanished slowly, beginning a few million years before the K/T Boundary. It could be, these scientists argued, that dinosaurs were on their way out long before any "killer rock" arrived from space.

Other scientists pointed out that dinosaur fossils are very, very rare—a

1

Dinosaurs flourished during the Cretaceous period, up until 65 million years ago. Their natural deaths during this time left bones that were covered by sediment.

2

The K/T impact may have killed all the dinosaurs and ended the Cretaceous period. The impact left a layer of debris made of material thrown out of the crater. This layer formed the K/T Boundary.

scientist might search his or her entire life and find only a few really good fossils. So it could be that a meteorite caused the dinosaurs to become extinct quickly, and their fossils haven't been found yet.

Over a long period of time, thousands and thousands of years, mud and sand pile up in rivers and streams or in the ocean, layer on top of layer. The oldest layer of mud and sand is at the bottom, and the youngest layer is at the top. In each layer are the fossilized bones of animals that died during the time that layer was formed. To find a dinosaur fossil, a paleontologist, a scientist who studies fossils, visits places where there are rocks dating from the time of the dinosaurs and begins to search.

Imagine you are a paleontologist working on a hill in the Hell Creek beds of Montana, a place famous for dinosaur fossils. Suppose the mud at the top of the hill is

This K/T Boundary sample from Montana is made of debris from a meteorite strike, so it looks different from the layers around it. You can clearly see the white K/T Boundary layer. The youngest dinosaur fossils yet known were found in Montana, about a meter below this layer. The dinosaurs that left those fossils probably lived only a few thousand years before the impact.

3

Slowly, over 65 million years, soil, rocks, and dirt piled on top of the layer left by the impact. Much later, bones beneath this layer were uncovered by erosion.

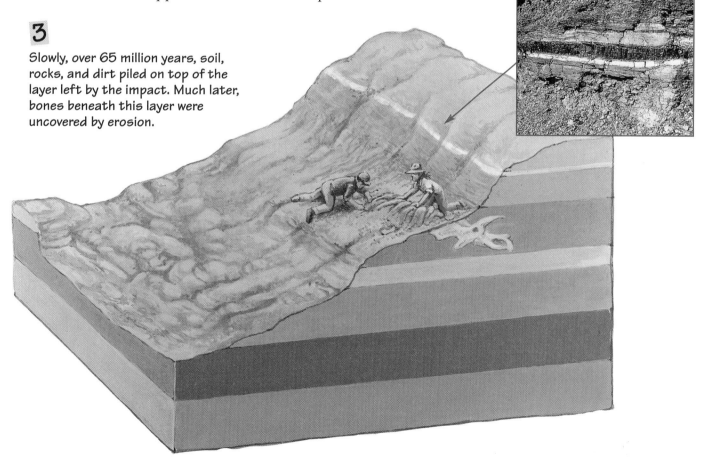

Evidence in the Earth

The force of a meteorite hitting the ground is enormous. Glassy solids called tektites can form from melted rock shot high into the atmosphere after a meteorite strikes. Tektites can be different shapes. Some look like spheres, some like discs, while others are shaped like rods or dumbbells. On average, tektites measure between less than 1 and 10 centimeters (about 1/3 to 4 inches) in diameter and weigh between 10 and 100 grams (about 1/3 to 3 1/2 ounces).

Tektites are found in K/T Boundary sites around the world, but the largest are within about 2,000 kilometers (about 1,200 miles) of the impact crater.

Tektites are usually glass, but tektites in the ocean near the K/T impact crater have turned into clay after sitting in seawater for millions of years.

from the K/T Boundary. Anything in that mud layer is about 65 million years old. All the layers below are older. As you walk from the top of the hill down to the bottom, you pass older and older layers of mud and sand.

After searching carefully, you find one dinosaur bone about halfway down the hill. Does that mean that the dinosaurs became extinct before the K/T Boundary? That could be right. Then, another paleontologist who's helping you finds a fossil nearer to the top of the hill. What does that mean? Clearly the dinosaur whose bone you found wasn't the last dinosaur. Did the other paleontologist uncover the remains of the last dinosaur? Do you have enough evidence to make a good guess about when the dinosaurs disappeared?

Most paleontologists agree that our record of dinosaur fossils is just too incomplete for us to know for certain when the last dinosaur disappeared. So we don't know if they all died at the K/T Boundary or not.

But since 1980 more and more evidence has supported Louis and Walter's meteorite theory. Much of this evidence is no bigger than a grain of sand. It includes microscopic diamonds, small bits of shocked quartz, and round or elongated glass pellets called tektites. Microscopic diamonds and shocked quartz can be made only under conditions of tremendous pressure, such as when a nuclear bomb explodes or a meteorite hits the ground. Tektites, some of which look like glass beads, form when rock is melted by the intense heat from a meteorite hitting the ground. Along with diamonds and shocked quartz, tektites are blown high into the air and then rain back to Earth. Together, these objects are called "impact debris." Impact debris forms layers in the earth, and scientists date this material using the fossils found above and below these layers.

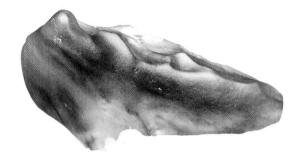

All these tektites are shown at their actual sizes. They were found all over the eastern hemisphere, from Australia and the Philippines to Vietnam.

The best evidence to support Louis and Walter's theory was the discovery in 1989 of an enormous meteorite impact crater called Chicxulub (pronounced "chick-sue-LUBE"). The crater is 180 kilometers (about 112 miles) in diameter and lies under the beaches of Mexico's Yucatán Peninsula. Tests have shown that the crater was made almost exactly 65 million years ago—the same time (give or take a couple of hundred thousand years) that the dinosaurs died. The crater is strong evidence that a massive meteorite strike happened at or about the same time as the K/T mass extinction.

But there still wasn't enough evidence that the impact happened at exactly the same time as the K/T mass extinction. Many scientists agree that events such as a giant meteorite impact and one of the largest mass extinctions of all time must have something to do with each other. Others argue that the impact, if there was one, happened after the extinction. It's very hard to tell who's right by looking at the fossil layers. Material is missing in the layers at many of the K/T Boundary sites discovered so far. That makes it difficult to be sure of the order of events. Not knowing the exact date of the impact is another problem. After all, a lot can happen in a couple of hundred thousand years!

This is a problem that interests many scientists, including myself. So, in 1994 I began to organize an expedition to uncover a more complete record of the mass extinction. The idea was simple, but it was hard to actually do. We would head out into the ocean on a ship designed for deep-ocean research and drill deep into the ocean floor. In the cores, or samples, of rock and mud retrieved from this drilling, we would look to see if the layer of impact debris—the diamonds, shocked quartz, and tektites from the meteorite—formed at the same time that ocean creatures became extinct. Other researchers had done this before, and had even drilled

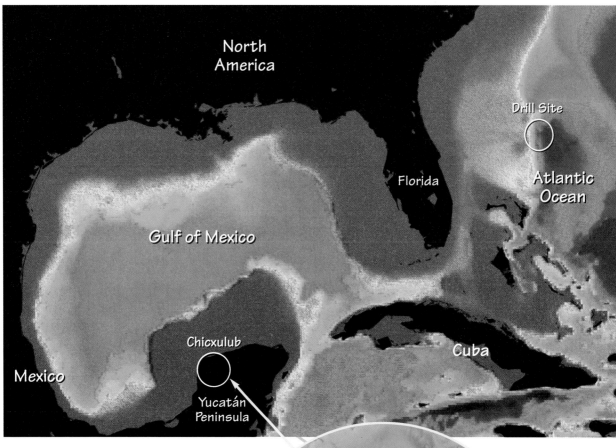

North
America

Drill Site

Florida

Atlantic
Ocean

Gulf of Mexico

Chicxulub

Cuba

Mexico

Yucatán
Peninsula

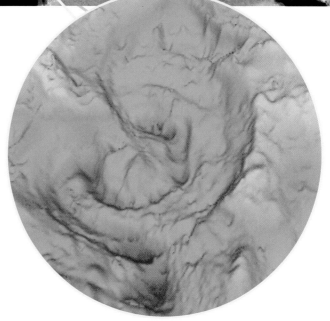

We chose a drill site off the Florida coast, a place about 1,900 kilometers (about 1,200 miles) away from the impact site off the coast of Mexico. If we drilled close to the crater, the force of the impact and the damage it caused would have likely jumbled the samples we were trying to find. So we chose a place away from the crater (shown in the satellite image above), hoping to find clear evidence of the meteorite strike.

Here I am on my way to look over the ship that will take me and a group of other scientists on a trip to gather information on the K/T Boundary impact.

close to our sites. But their equipment wasn't as good as ours, and they failed to recover undisturbed evidence of the K/T Boundary. The site we chose was 1,900 kilometers (about 1,200 miles) from the impact crater. We wouldn't be able to answer for sure the question of how the dinosaurs died. But if we found that the meteorite struck at the same time the extinction of ocean creatures occurred, we would have strong evidence that a meteorite impact had caused a disaster that could have killed the dinosaurs, too.

We would be detectives trying to find a killer. We had plenty of dead bodies, namely all the dead dinosaurs and other extinct forms of life. We also had a crater in Mexico that was about the right age. But was the crater blasted into Earth at precisely the same time that up to seventy percent of all species died? We had clues, but we were missing a key piece of evidence—the certain knowledge that a meteorite impact and a mass extinction happened at the same time.

OFF TO FIND MORE CLUES

"It was as though the creatures had drunk a magic potion, shrinking them all to an eighth of their former size."—Richard Norris

(above)
Ocean drilling can be expensive. This is scientist Mary Anne Holmes from the cruise, sitting by a broken drill bit. Drill bits are made to be durable, but they can break. This bit alone costs about seven thousand dollars.

(left)
The JOIDES Resolution is run by the Ocean Drilling Program, an international scientific organization. Scientists come from all over the world to work together aboard the ship, and many become great friends.

It was early January 1997, about 500 kilometers (about 300 miles) off the coast of Florida. I wondered what had happened to the tropical weather as I pulled on my jacket at about dawn. A winter storm was coming. The sea around our research drilling ship, JOIDES (Joint Oceanographic Institutions Drilling Expeditions) *Resolution*, was dull gray and speckled with whitecaps. Still, the ship hardly moved. It was more than 145 meters (about 475 feet) long, with sixty crew members and scientists, plus laboratories, cabins, a helicopter pad, a drilling derrick, and space for 10 kilometers (about 6 miles) of drill pipe. It would have taken a real storm to rock the ship.

Winches whined as the drillers brought up new samples from the bottom of the ocean. One of the scientists on board, Jan Smit, had studied the K/T Boundary nearly everywhere in the world. He had also studied rocks and mud collected during a past cruise to this same place in the early 1980s. Those samples held small amounts of tektites and fossils of both the late Cretaceous and the earliest part of the Paleogene periods. Unfortunately the fossils and tektites were hopelessly mixed together by the drilling equipment that was used at the time.

21

Drilling in the Deep Sea

Drilling from a ship in the ocean is a complicated job. About 5,000 meters (about 16,000 feet) of hollow drill pipe need to be assembled and lowered to the ocean bottom. The hollow pipe, called a drill string, is lowered from the derrick on the ship's deck. Cores of seafloor material as long as 9.5 meters (about 30 feet) are collected in plastic tubes called core liners that are inside a device called a core barrel. The core barrel is pulled up to the ship by a wire cable inside the drill string.

One of the challenges of open-ocean drilling is working on a ship that is constantly moving. To help keep the drill string straight, the pipe has a device called an acoustic beacon that sends out a signal. This signal is monitored and the ship's position is adjusted to keep the ship directly above the drill site. When there is a storm or high waves, it's sometimes necessary to leave before drilling is finished. If this happens, the ship leaves behind a device called a reentry cone so that it can drill in the exact location later.

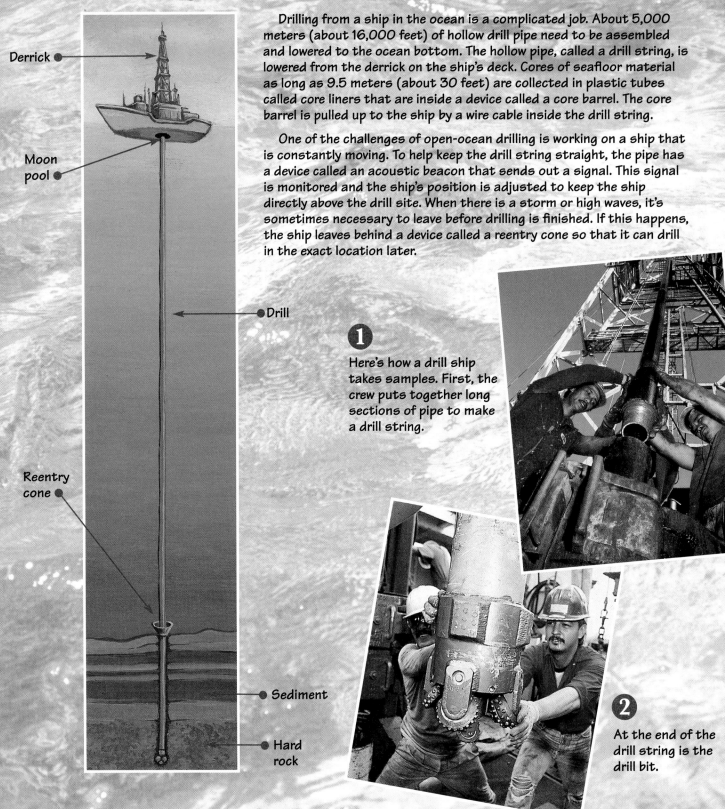

Derrick

Moon pool

Drill

Reentry cone

Sediment

Hard rock

1 Here's how a drill ship takes samples. First, the crew puts together long sections of pipe to make a drill string.

2 At the end of the drill string is the drill bit.

③

The drill string is lowered
in sections through a hole
in the bottom of the ship
called the "moon pool."
After hours of lowering,
the drill string reaches
the ocean floor and drilling
can begin.

④

To drill, the "piston corer,"
a kind of giant nail gun,
shoots the core barrel
into the seafloor from
inside the drill string.
After drilling, the core
barrel is brought back
on deck. Here, the core,
safe in the plastic core
liner, is being removed.

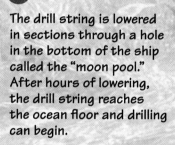

⑤

After the core is taken
from the core barrel (far
left), a new liner is added
to the core barrel. The core
barrel is sent down and
brought up over and over
until drilling is finished.
The drill string is then
brought back on deck and
stored until the next
drilling site is reached.

6

Next, the core is cut into smaller lengths to make it easier to handle.

 7

Splitting a core in half (lengthwise) is exacting work. Once the cores are sliced, half of each core is labeled and stored. No samples will be taken from one half so there will always be an intact core sample to use as a reference.

 8

Cores that have not yet been sliced are labeled and temporarily stored on racks.

The mixing of sediment caused by old drilling equipment made it impossible to tell what happened when. Our equipment was much better and would let us see the sequence of the impact debris layer and evidence of the mass extinction of life in the ocean. There was also a good chance we'd recover a clear record of the sequence of events during the impact itself.

We stood around on deck, watching the sky lighten with the coming of dawn. The drillers stored the pipe or "core" filled with mud they had just pulled up from the seafloor. They then began to drill a new core under the glare of spotlights that shine all night long. Only when drilling started again did the chief driller pick up the first mud-filled core. He unscrewed the drill bit from the end of the core and slid the clear plastic core liner full of mud out of its metal case. You could see through the liner to the mud inside. Jan said we should look for a light-green layer of impact debris about 2 or 3 centimeters (about 1 inch) thick.

We were almost 1,900 kilometers (more than 1,000 miles) away from the crater, so we didn't expect the impact layer to be very thick. Most of the largest rocks and pieces of debris should have fallen in the ocean much closer to the crater. It's been hard to find samples close to the crater in which the sequence of the layers is easy to see. Researchers think this might be because the giant waves kicked up by the impact would have tumbled and mixed up the layers. By drilling out in the Atlantic Ocean, we had a better chance of finding undisturbed layers of debris.

Suddenly Jan let out a whoop. After a quick look at the core, we all joined in, jumping around on deck and slapping each other on the back. There, in the middle of the core, was the green layer we had all hoped to find. But it was not 2 or 3 centimeters thick—the layer was at least 17 centimeters (about 7 inches) thick! How could so

Getting the Color Right

Scientists match the colors of the sediments in the core to those in a reference book called a Munsell Color Book. That way, other scientists can know exactly what color the sample is without having to look at the core itself.

The scientists also look at thin slices of the core under a microscope.

glass microscope slides ●

knife to slice off tiny bits of core ●

core samples ●

Munsell Color Book ●

fold-up magnifying glass

small pick used to take samples

Jan Smit was the first person to examine the K/T Boundary in the core we pulled up from the bottom of the ocean.

much debris be thrown so far from the crater? Thinking about it, we decided that the layer we found may have come from debris, or ejecta, thrown out of the crater in "rays," like the streaks around some impact craters on the moon.

Excited by our find, we carried the core into the core-splitting room. Paula Weiss and Gus Gustafson, both veterans of many research cruises, used a special saw to slice the core in half. Most of the core was full of light-gray mud, looking like pottery clay. But in the middle was a bright green-blue layer of material that looked like fine beach sand. It was topped by a razor-thin line of rust and a dark-gray band about 4 centimeters (1.5 inches) thick.

Later we would find out that the green layer was made mostly of tektites. The tektites' shapes suggested they were probably formed from melted rock that hardened as globs of it spun through the air. Bits of chalk, limestone, granite, and shocked quartz were also in the green layer. All these kinds of rock have been found in the Chicxulub crater.

Debris from a meteorite impact is sometimes thrown out in rays, like these in this illustration of a meteorite crater on Earth.

Crater: a circular depression in a surface caused by an impact

Ejecta: material shot out of a crater, often forming rays

Floor: the interior of a crater

Rim: the raised edge of the crater formed by the outward and upward compression, or squeezing, of the crater walls and ejecta—on a large crater, this rim collapses

The dark-gray layer above the green tektite layer contained a large amount of a metal called iridium. Iridium and its close cousin, platinum, are common in meteorites but very rare on Earth's surface. It was the high amount of iridium in rocks dating from the mass extinction that originally led Louis and Walter Alvarez to their meteorite impact theory. Only a large meteorite impact could explain so much iridium. So, if the dark-gray layer came from a meteorite, it was very likely that the green layer next to it was made of debris from that meteorite's impact. We had another piece of the puzzle. But was this meteorite impact strong enough to kill the dinosaurs? I hurried down to the ship's paleontology lab to find out what the core could tell us.

Brian Huber, a paleontologist from the Smithsonian Institution in Washington, D.C., was hunched over his microscope. He was studying the tiny fossils from the mud above and below the green tektite layer. Brian pushed back his chair, grinned, and said, "You have got to look at this!" Seen through a microscope, what had looked like table salt to the naked eye turned into little shells. Some looked like popcorn. Others looked like

Cores are studied in several different ways. One way is to take samples of the cores. Here, Ocean Drilling Program technician Tim Fulton (left) and scientist Brian Taylor of the University of Hawaii are hammering small plastic tubes into the core to take small "core samples" to examine. Researchers then study these samples under a microscope. You can see a green debris layer in the middle core (inset).

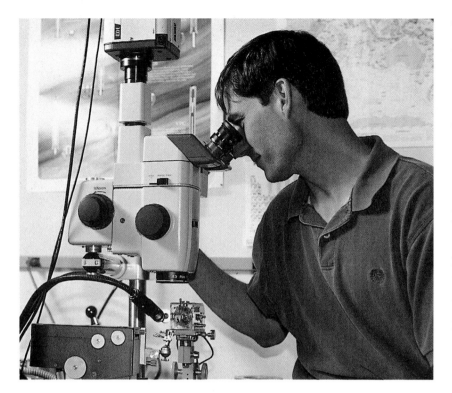

Brian Huber studies microscopic foraminifera so small that they need to be magnified fifty to one hundred times to be seen.

This is a tiny foraminiferan shell resting on a person's finger. At right, you can see what these shells look like under a microscope.

flying saucers covered with little bumps, spines, and ridges. These were the fossil remains from the layers that had formed just before the meteorite hit the ground. I was looking at shells of foraminifera (pronounced "fo-ram-in-NIF-ahr-uh"), tiny, single-celled creatures that have floated in the oceans by the trillions since the time when the dinosaurs roamed the earth.

Brian slid a second sample under the microscope, and I looked again. This sample was from just above the tektite layer, showing the fossils of creatures after the meteorite impact. The sample looked like something from another planet. The fossils were like little bits of dust, even under the powerful microscope. It was as though the creatures had drunk a magic potion, shrinking them all to an eighth of their former size.

But there was something else. The great variety of fossil shapes was gone. There were only three or four different kinds of simple, nearly smooth shapes. Here was solid evidence of a mass extinction in the ocean happening right after the meteorite impact. Some seventy different species of these tiny creatures lived in the oceans before the impact, but only three or four were alive after the impact.

At last we had dramatic evidence that the meteorite impact had caused a disaster in the sea.

Tiny Signs

One of the best records of the mass extinction comes from deep-sea microfossils, skeletons of tiny animals and plantlike organisms. One group of organisms with an extensive fossil record is foraminifera, whose shells look like table salt to the naked eye.

Foraminifera shells are widely used to study ancient oceans. Scientists can learn many things from studying their shells, such as the water temperatures of the seas in which the foraminifera lived.

Above are the shells of foraminifera that lived before the K/T Boundary. The shells are elaborate, and they have many different shapes.

Below are samples of the shells of foraminifera that lived after the K/T Boundary. These shells are much simpler than those that existed before the K/T Boundary. Clearly something dramatic had occurred.

4

IT CAME FROM THE SKIES

"There have been many impacts, but they haven't all caused death and destruction." —Geologist Peter Schultz

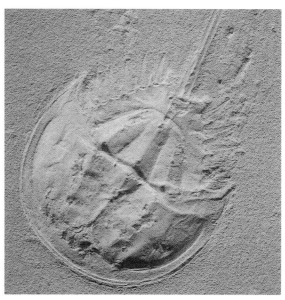

Good as they were, the new cores we pulled up far out at sea that day didn't answer all of our questions. The cores showed how tiny floating ocean creatures responded to the meteorite strike. But they didn't show what happened to larger sea creatures. And they didn't show what happened to animals and plants living on dry land.

These are pieces of the puzzle that scientists have to sort out by inference—making educated guesses based on the evidence we have. Our deep-sea core showed that the impact event drastically reduced the number of microscopic plants and animals in the oceans. From that we can infer that other animals also suffered, because ocean life depends on these tiny animals and plants for survival.

There are parts of the story we may never be able to answer completely. For instance, ocean cores don't tell all the details of how the impact happened or how long it took for animals to begin dying around the world. Sediment, made of mud and sand, piles up very slowly on the seafloor, and the record it preserves is often incomplete. It usually takes a thousand years or so for even a

(above)
Some sea creatures survived the K/T extinction. These survivors included the horseshoe crab. This horseshoe crab fossil was found preserved in a fossilized lagoon in southern Germany inside a layer of limestone from the Jurassic period.

(left)
Other sea creatures weren't as lucky. The ammonite, an animal similar to an octopus and once numerous in the ocean, became extinct at the end of the Cretaceous period. This ammonite fossil was found in England.

31

A Closer Look

This photograph of a section of a seafloor core shows about 100,000 years from the end of the Cretaceous period to the beginning of the Paleogene period.

We can roughly tell from the layers the time when events happened, but it's not possible to be exact. However, information from cores like these can help scientists piece together long-ago events.

couple of centimeters of mud to pile up on the seafloor. Because of this, the layers show just pieces of events rather than what happened in the hours, weeks, or even hundreds of years after an event. Also, the pieces themselves may be jumbled, since burrowing animals may have mixed them up. So, determining exactly what happened during the K/T event from a core sample is like reconstructing 2,000 years of history using only jumbled bits of information from the years 1, 1050, and 1930. Information on the Roman Empire, the rulers of China, and the Renaissance might be there, but it would all be hopelessly mixed together. It would be almost impossible to reconstruct the events in any detail.

This is the end of the Cretaceous period. The mud contains a diverse sampling of the microscopic shells of plants and animals typical of the oceans during the Cretaceous period.

This is the beginning of the K/T Boundary layer that separates the Cretaceous and Paleogene periods.

This layer of impact debris from the K/T meteorite strike includes glassy green tektites and bits of rock that were blasted about 1,900 kilometers (more than 1,000 miles) away from the impact crater.

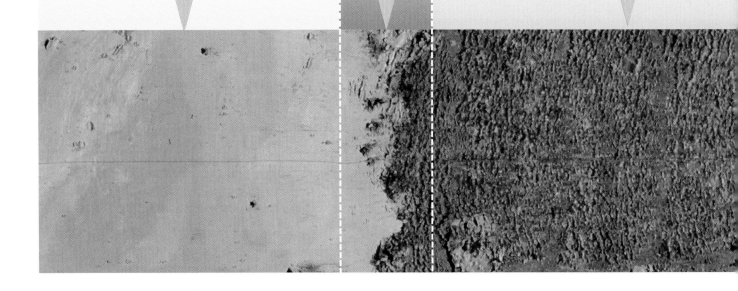

So, how do we figure out how the K/T event happened in the very, very distant past of geologic history? One way is to use laboratory experiments and computers. For instance, some scientists study impact craters on the moon, while others study large volcanic eruptions to understand how dust and debris move in Earth's atmosphere. Some scientists study and conduct experiments on the meteorites themselves. Other studies use computers to simulate, or make models of, events that happen too quickly or too rarely to observe easily. The results of experiments and simulations can be very useful in understanding what may have happened during and soon after the meteorite impact on Earth.

This is the end of the K/T Boundary layer. After this layer is the beginning of the Paleogene period.

This shows a decrease in ocean plant and animal life after the meteorite impact. It also shows the fine dust and debris from the impact.

Here is the beginning of the repopulation of empty seas. New life is appearing from survivors of the K/T extinction.

The amount of life on Earth begins to return to pre-impact levels, but this process is not complete for about ten million years.

Asteroids and Comets

Asteroids are generally composed of rock, with some made of metals such as iron and nickel. Asteroids that strike Earth come from the asteroid belt between Mars and Jupiter.

On the other hand, comets like Hale-Bopp, above, come from orbits beyond the outermost planets and are made of rock and ice. Comet Hale-Bopp has twin tails that are more than 150 million kilometers (almost 100 million miles) long. The small bright streak to the right of the comet is a meteor.

Judging from the speed of modern comets and asteroids, a rock on its way to Earth would probably be moving at a speed of between 15 and 72 kilometers (about 9 to 45 miles) per second. This is much faster than a speeding locomotive (moving at a mere 0.04 kilometers per second) or even a speeding bullet (about 1 kilometer per second). The speed is critical, because the explosive power of an impact increases dramatically with the speed and size of the object making impact.

A rock the size of a minivan hitting Earth could cause an explosion almost equal to the nuclear bomb that was dropped on Hiroshima. A meteorite the size of a house could unleash an explosion more than a hundred times more powerful. The force of the explosion could be about equal to the detonation of one of the largest nuclear bombs ever made. Increase the meteorite size to a kilometer or so—a little over half a mile across—and sling it at Earth at twenty kilometers a second, and you could have an explosive power greater than that of all the nuclear bombs on the planet.

This meteorite landed in Whitley County, Kentucky, in 1919. It weighs more than 14 kilograms (31 pounds).

The object that hit Earth 65 million years ago was perhaps ten kilometers (about six miles) across with explosive power more than 10,000 times the power of all the world's nuclear weapons. No wonder there was a mass extinction in the ocean. It's surprising that anything survived at all!

Just what was it that hit Earth 65 million years ago—an asteroid or a comet? This turns out to be an important question, because asteroids and comets travel at different speeds. That means they would have very different explosive forces when they hit a planet.

Most asteroids orbit the sun in the same direction as Earth does, and at nearly the same speed. So an asteroid would hit Earth at only about 20 kilometers per second, or about 12 miles per second. A comet, on the other hand, could zip in from the outer reaches of the solar system and hit Earth head-on at speeds of up to 80 kilometers (almost 50 miles) a second. It would be as different as being hit by a car traveling in the same direction and moving only a little faster than you are and being hit head-on by a car moving the wrong direction on a freeway. The first crash would dent your bumper, but the second would destroy your car.

Current thinking is that the K/T meteorite was an asteroid, because bits of rock that may have come from an asteroid have been found in K/T Boundary layers. How do we find out? One way to study large meteorite impacts is to use computers. Computer simulations may not always be accurate (it depends on how well simulations are set up and the data that are entered), but they can be helpful in figuring out what might have happened and why.

Cross sections of meteorites can provide scientists with important clues about other meteorites that have landed on Earth.

After a 1960 fireball was seen in the Wiluna District of Western Australia, many stones like this one were found. This stone weighs 28.3 grams (about 1 ounce).

After four explosions in 1924 over Weld County, Colorado, 27 stones like this one were found. The largest stone weighs 23.5 kilograms (about 53 pounds).

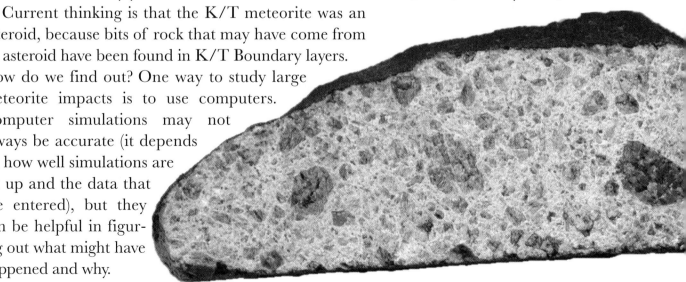

Modeling Meteorites

In this laboratory experiment, a small glass pellet is shot into a sand target. Scientists can use what they learn from studying small impacts like this to help them understand a large impact. This pellet was fired at an angle to the target, similar to the way the K/T meteorite might have hit the earth. The angle is important, because a low-angle impact like this one spreads more debris and hot gas into the atmosphere than does a hit directly from above. You can see a wave of debris material fanning out to the left of the impact site below.

Scientist Peter Schultz says, "It's important to know the angle at which [the impact] happened. The lower the angle, the more destruction that may have occurred. There have been many impacts, but they haven't all caused death and destruction. From modeling impacts, researchers have found that a wave of low-angle material creates a 'corridor of destruction.' It may have been just such a corridor of destruction that contributed to the end of the dinosaurs."

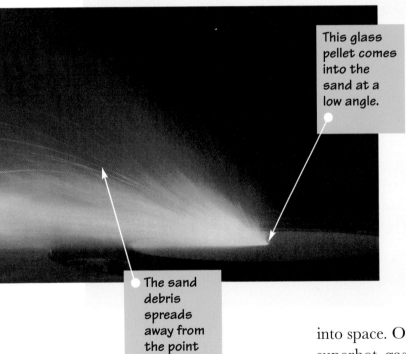

This glass pellet comes into the sand at a low angle.

The sand debris spreads away from the point of impact.

We can also infer what happened after the impact by making our own "impact event" in a laboratory and then watching what happens.

One person who does this is Peter Schultz of Brown University in Rhode Island. Peter uses a hypervelocity cannon at NASA's (National Aeronautics and Space Administration's) Ames Research Center in Sunnyvale, California. The cannon fires a marble or a glass pellet at extremely high speeds. Peter's cannon blasts special targets, and he films the results using high-speed cameras.

According to Peter, in a big impact, more than two-thirds of a meteorite is melted and vaporized —turned into dust or a fine spray. The rest of the meteorite is shattered into sand and gravel-sized chunks. The place where the meteorite strikes the ground is also vaporized, blasting a crater much larger than the original meteorite. What happens to all the smashed-up debris?

Peter's work suggests that some of the vaporized meteorite and debris might simply shoot back into space. Other debris could join vapor to form a jet of superhot gas that would rocket along the ground at supersonic speed. This supersonic gas-jet might also blast out of the atmosphere, carrying a mass of burning trees,

The vacuum chamber allows simulations of conditions on other planets. Large impact craters are preserved on other worlds, but are destroyed by erosion on Earth. By comparing craters produced by the gun, where the conditions can be controlled, with craters on other planets, scientists can understand what happens on Earth.

The projectile is shot through a long barrel at speeds up to twenty times faster than an airplane flies. When in place vertically, the gun is almost three stories high.

The elevating screw is used to lift and lower the gun arm to different angles.

Large windows on each side of the chamber allow scientists to record what happens with high-speed cameras and instruments.

soil, and boiling water with it. While this is happening, one last part of the cloud of hot vapor from the impact might rise above the crater, cool slightly, and form tektites. The tektites would return to Earth as a rain of glass a few minutes to a few hours after impact.

Peter's experiments, added to computer simulations and studies of tektites in K/T Boundary layers, paint a picture of awesome destruction. Much of the initial destruction from the K/T meteorite was probably over in less than an hour, but some effects would be felt for tens of millions of years. So how might this disaster have unfolded?

This is NASA's Ames Research Center Vertical Gun Range. The hypervelocity cannon shown above is used to fire material at targets. Scientists use it to find out how speed, angle of impact, and the material of both the "meteorite" and the target affect the way debris is spread.

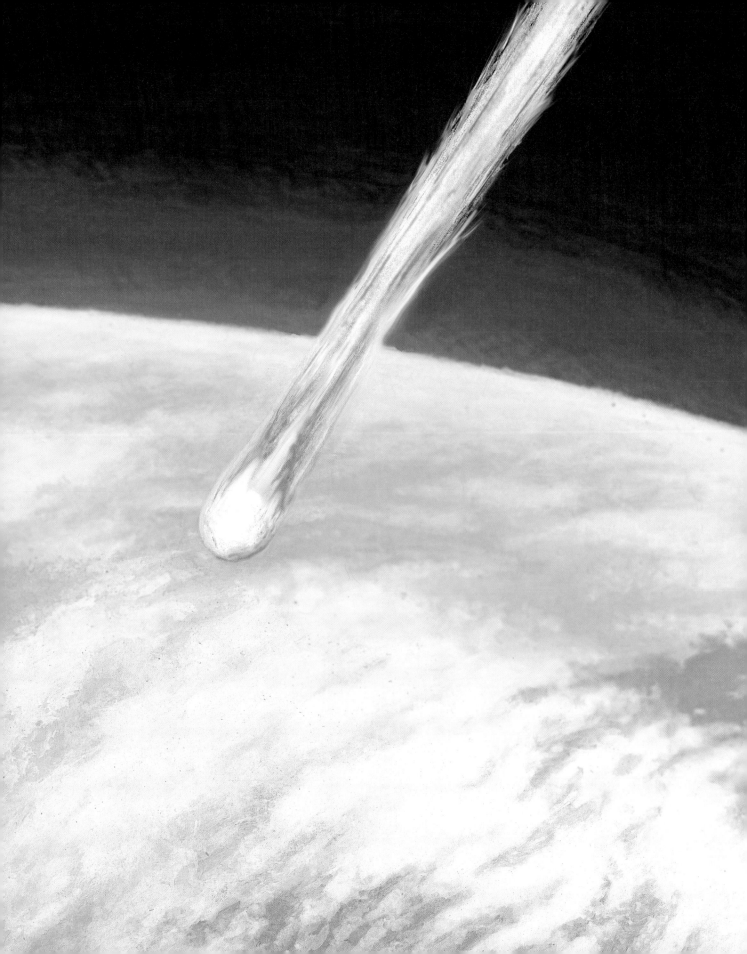

5

THE END OF THE DINOSAURS

"With nowhere to hide, the dinosaurs would have all died."—Richard Norris

It's possible that the dinosaurs saw the meteor long before it hit. It may have been a beautiful comet with a sweeping tail that grew larger and brighter over several months. Or it may have been an asteroid tumbling through space on its way to Earth. No one is certain. In either case, once the meteoroid punched through the atmosphere, it reached the ground in only a couple of seconds.

Even before the meteoroid reached the ground, it would have done a great deal of damage. First, the hole in the atmosphere would have caused supersonic winds to form as air rushed to fill the gap. This would have caused a flash of light and a rolling sonic boom. Second, the heat of the swiftly falling meteoroid would have roasted the atmosphere, turning the nitrogen in the air into chemicals that form powerful acids. Later this acid would rain down on the earth, destroying plant life wherever it fell.

As the meteoroid bore down on the Yucatán Peninsula in Mexico, it would have squashed the air beneath it, but it would have been moving too fast for all of the air to escape. The air that was trapped between the meteoroid and the ground would have been pushed ahead of the meteoroid like a giant battering ram. Then, the air would have been forced into the shallow sea that covered the Yucatán, and from there into the rock at the bottom of the ocean.

(above)
Astronomers at the Spacewatch Project in Tucson, Arizona, use huge telescopes to identify near-Earth objects that may threaten our planet. But it's estimated that a meteor impact like the one from the K/T period happens only once every 100 million years.

(left)
Whether a comet or an asteroid, a large meteoroid moving through Earth's atmosphere would have a tremendous amount of speed and explosive force as it neared the ground. This picture shows a meteoroid heading toward Earth to strike at a nearly vertical angle. The K/T meteoroid may have struck the earth at a lower angle.

39

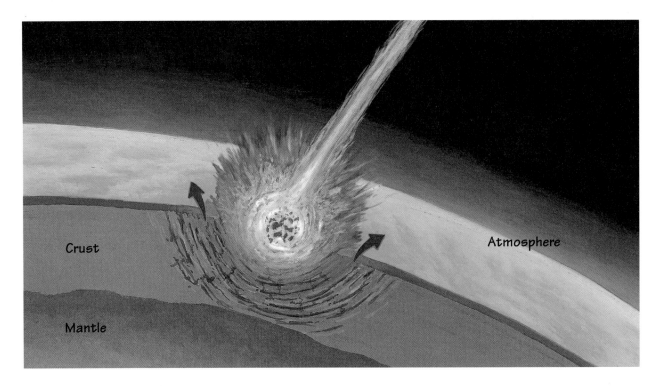

Crust

Mantle

Atmosphere

The meteoroid would be moving so fast that a solid mass of air would pile up in front of it. The air would actually hit the ground first and dig out a crater.

As it moved closer to Earth, the meteoroid's speed would have caused air and water to act like solids. A similar thing happens when you jump into a pool and do a belly flop. It hurts mostly because the water can't squeeze away from your body fast enough. It feels as if you have landed on solid ground. The meteorite would have shattered, melted, and even vaporized as it penetrated deep into Earth in less than a second. Even water would have seemed like rock.

A few seconds after impact, the crater would have been more than 40 kilometers (almost 25 miles) deep—more than four and a half times the height of Mt. Everest. Eventually the center of the crater would have risen in a dome shape. This shape is formed after a large impact because the rock in a crater floor recoils like a giant spring as the rim of the crater collapses inward. Rock melted by the impact would have covered the crater floor. The heat of the impact may have turned the ocean water near the crater into steam.

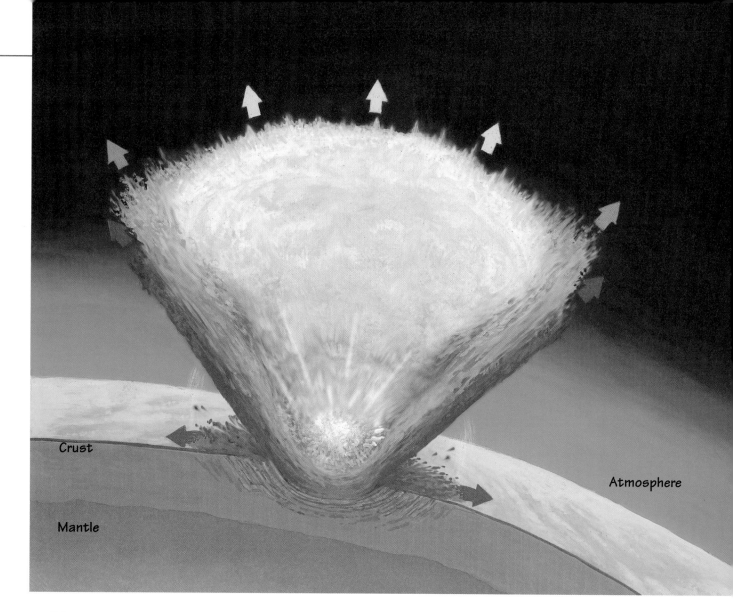

Crust

Mantle

Atmosphere

The air and water squeezed by the meteorite would have mixed with the steam and the melted rock in the crater to become a huge plume of hot gas and ejecta. This plume would then have exploded out of the crater because of the high pressures and temperatures created by the impact. Most of it would have blown back up through the atmosphere and into space. Some of it would have cooled as it reached the outer edges of the atmosphere and formed tektites that rained back to Earth over great distances.

If the meteorite struck Earth at a low angle to the surface (about 30 degrees), the fireball would have blasted

The meteorite breaks apart. Ejecta is shot from the crater, which now extends halfway through Earth's crust. The yellow arrows show the trajectory, or path, of ejecta and expanding vapor. The debris spreads far over the earth's surface. The tan arrows show the path of melted debris that shoots above the atmosphere. This rains down later as tektites. The lower path of ejecta is marked by the red arrows. The blue arrows show the path of ejecta that join a ground-level, supersonic blast of water vapor and hot gases.

Large animals like these hadrosaurs would have had no defense against the results of a large meteorite impact.

away from the crater at ground level heading north across the Americas. Because destruction from the impact seems to have been worse north of the crater, scientists think the meteoroid may have come from the south. Peter Schultz and fellow scientist Steve D'Hondt think the meteoroid may have sailed over South America and then hit the Yucatán at an angle, causing part of the fireball to shoot off at a low, northerly angle.

This fireball of dust and debris would have been powered by superheated wind and traveled faster than a jet airplane. It would have reached Montana in about ten minutes, ripping up soil and trees as it passed. Dinosaurs would have had little warning, because the fireball would have arrived before even the impact's sonic boom.

Imagine a herd of hadrosaurs near what is now Montana. The lumbering beasts would have looked up to see a dirty-orange curtain of trees, mud, and dust coming from the south. There would have been nowhere to run as the blast picked them up like dry leaves or knocked them flat and killed them with its heat. Nearly all of

North America would have been a blackened wasteland within twenty minutes after the meteorite hit.

Animals in low-lying land between mountains may have survived the fireball that scoured the earth, but they might have faced an equal danger. Tektites would have begun to "rain" back down to Earth. The rain of tektites must have been a beautiful but deadly sight 65 million years ago. After the giant meteorite struck, the trillions of tektites falling to the earth's surface would have lit the night sky in a blaze bright enough to turn darkness to daylight. And each spark of light would have heated the sky a small bit. Add up the trillions of streaks of fire, and the heat would have been tremendous. With nowhere to hide, the dinosaurs would have all died. Forests would have burned, too. A huge wildfire would have swept much of the globe, sending enough smoke and soot into the air to blot out the sun.

But the destruction would not have ended there. The Yucatán Peninsula was covered by a shallow ocean at that time, and a meteorite crashing there would have been like dropping a rock into a puddle. Some of the water would have been vaporized or blown into space, but much of it would have formed giant ripples that moved outward. These ripples are called tidal waves, but are more properly known by the Japanese name *tsunami*, or "harbor wave." Tsunamis look like ripples in the open ocean, but when they enter a harbor or other shallow place, they build to enormous heights and sweep inland for miles. The meteorite strike could have caused tsunamis more than 100 meters (about 300 feet) high.

Mud from cores taken in waters off Florida, the Gulf of Mexico, and the Caribbean also show signs of huge landslides started by the impact. Mud deposited before the impact shows evidence of a great upheaval in the ocean. The layers are twisted, as if the mud slid down in a giant landslide.

A tsunami may seem small on the open ocean, but it can turn into a gigantic wall of water, causing widespread destruction as it hits the shore. A tsunami that's only twenty centimeters (about eight inches) high on the open ocean can form a wave ten meters (about thirty feet) high in shallow water. The coast of Lituya Bay in Alaska (above) was stripped of life by a tsunami that hit on July 10, 1958. All of the brown areas are places that were covered with trees and grass before the tsunami hit.

6

THE LONG NIGHT

"The K/T meteorite impact could have thrown the whole planet into blackest nighttime."—Richard Norris

The meteorite impact's most deadly effects were in North America and the Caribbean. So why didn't the dinosaurs survive in other places? One reason may have been the rock in the crater.

Much of the rock where the meteorite hit, and parts of the meteorite itself, probably became tektites. But other parts were blasted into bits so tiny that a bucket of this "dust" would pour like oil. Computer experiments suggest that the impact could have blasted enough dust into the atmosphere to blot out the sun for weeks, even months. The K/T meteorite impact could have thrown the whole planet into blackest nighttime—total darkness without the light of a single star. How many days could animals have lived without food while they waited for the dust to be washed from the sky?

As if all of this weren't enough, the Yucatán Peninsula itself may have provided the final blow for much of life on Earth. The Yucatán rests on sulfur-rich rocks. When sulfur burns, it creates sulfur dioxide—a main ingredient in sulfuric acid. The heat of the impact could have created huge amounts of sulfuric acid and sent it

(above)
In 1991 Mt. Pinatubo, a volcano in the Philippines, erupted and blasted so much sulfuric acid into the skies that Earth cooled slightly for about three years afterward.

(left)
Ash and steam from Mt. Pinatubo rose in big clouds over a portion of U.S. Clark Air Force Base in the northern Philippines. About one thousand people working at the base had to be evacuated.

45

Disasters in one place can affect distant places. In 1998 smoke from fires in Mexico and Central America blanketed the land all the way north to Texas. Smoke plumes blew as far north as Wisconsin. The path of the white smoke is visible on the satellite image above.

into the sky. Sulfuric acid is poisonous, and the acid rain it creates is hard on animals and plants.

And there's one more thing about sulfuric acid—it reflects sunlight. The rock dust, sulfuric acid, and water vapor in the atmosphere would have kept a lot of sunlight from reaching the ground. Without sunlight, plants would have had trouble growing. Even worse, without sunlight, Earth would have suddenly grown a lot colder.

This cooling would have been more extreme than anything we have ever seen. Computer programs suggest that a big meteorite impact could throw the centers of the continents into a deep freeze for perhaps a year. Because water stores more heat than soil, the land near the water would have stayed relatively warm compared to the frigid continental centers. In the ocean the major problem would have been the lack of sunlight until the dust in the atmosphere cleared, not the cold temperatures. Just like land plants, ocean plants would have needed sunlight to grow. Their failure to grow would have affected everything depending on them for food.

And problems wouldn't have ended with turning out the lights. The burning meteorite's entry into the atmosphere plus the tektite shower would have destroyed Earth's ozone shield. Ozone is a gas that soaks up the sun's harmful rays, those that cause sunburns.

Not only would there have been lots of sunburned creatures walking around after the dust cleared, but mutants, or living creatures that have experienced a change in their genes, may also have stalked the land. The same harmful rays that cause sunburns can also cause mutations, or changes, in animals' cells. The destruction of the ozone shield might have increased the rate of cancer caused by mutated cells. But it may also have helped changes to take place that allowed other animals to survive.

Once in a while, a change, or mutation, may permit an animal or plant to do something new, like run a little faster, pick up things with its paws, or see better in the dark. Such changes might be just what an animal needs to find more food or to be able to avoid predators more successfully. Some of the more useful mutations helped descendants of animals that lived through the K/T meteorite impact survive.

Sargassum fish have developed the ability to blend with their surroundings. Different fish may be different colors, but they always match their own weed patches. If caught by a predator, these fish also have the ability to fill their bodies with water until they become too large to be swallowed. These useful mutations help Sargassum fish survive.

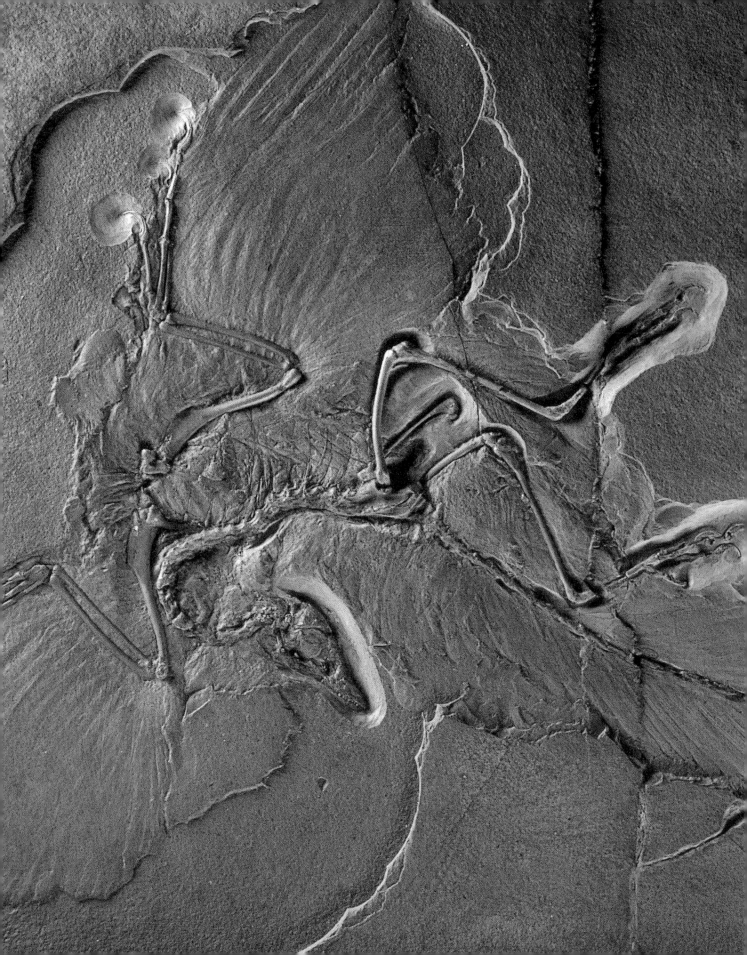

7

SURVIVORS

"I find it amazing that the extinction was not more widespread."—Richard Norris

The K/T extinction was a deadly mix of an impact blast, fires started by falling tektites, acid rain, blockage of the sun caused by dust, and the destruction of the ozone shield. I find it amazing that the extinction was not more widespread than it was.

Plenty of animals and plants died, but in many cases there were enough survivors to start life over again. At least thirty percent of Earth's species survived.

Why some animals made it and some didn't is one of the mysteries of the K/T extinction. We don't know for sure why most survivors thrived, but we can make some guesses. Luck was probably part of it—the luck of being in a bit of forest that didn't burn, or in a burrow when the shock wave and fireball rolled overhead. Another reason could have been adaptability. Animals able to eat a lot of different kinds of foods had more of a chance of finding something they could eat than those restricted to one kind of plant or one type of prey. Survival may also have depended on where an animal lived or whether it had the right skills to live through a time of little food.

For instance, deep-sea animals live in a world where

(above)
Some animals survived the K/T disaster, including the ancestors of modern birds such as this crow.

(left)
One bird ancestor might have been the archaeopteryx (or "ancient wing"). This fossil dates from the Jurassic period. The archaeopteryx is part bird and part dinosaur. Unlike modern birds, it had teeth, three claws on each wing, belly ribs, and a long, bony tail. But it also had feathers and hollow bones. The archaeopteryx was about the size of a modern crow and might have been able to fly, but not very well.

The triceratops (or "three-horned faced") dinosaur looked menacing but ate only plants. The armor on its head was used to attract mates and to fight with rival males. The armor was probably also useful for scaring away potential predators, though at a weight of six tons and a length of nine meters (about thirty feet), triceratops was probably safe from other animals. Nevertheless, its impressive size and armor didn't protect it from the K/T extinction. Triceratops, along with the other dinosaurs, didn't survive.

nearly all their food sinks down from above. For many creatures here, there is a lot of food once a year when the ocean surface blooms with life. That life then dies and sinks to the animals waiting below. These animals are used to eating nonstop for a few weeks, then more or less "turning off" to wait for more food from above. The mass extinction would have provided extra food for animals in the deep ocean, at least for a while. When the food was gone, the deep-sea creatures would have entered a kind of sleep.

For others in the deep seas, food comes from the earth itself, in chemicals oozing out of vents in the seafloor. For these animals, the death of much of the surface world would not have mattered at all.

Some microscopic floating plants were able to seal themselves into their shells until the bad times were over. Other kinds of microscopic plants died by the trillions and took with them much of the rest of the food chain. What do you, as a plant eater, do if all the plants on Earth die? More to the point, what do you do as a five-ton triceratops when all the trees and shrubs you eat are dead? The answer is simple—you also die.

Still, there were surprising survivors. Sharks and shore birds made it. Turtles, crocodiles, snakes, and our very distant mouselike relatives all lived through the extinction. Dragonflies and bugs of many different kinds hung on as well.

The survival of many of these animals is understandable. Many were already used to sleeping away at least part of the year and could hibernate. Plants might have survived as seeds or bulbs that could wait years to sprout. In one form or another, many of the survivors slept.

But how did sharks and birds make it through the disaster? Many sharks could have gone to sleep, and unborn offspring could have survived in egg cases—but birds? Bird fossils are rare. The fossils we have managed to find of birds that survived the K/T event usually belong to shore birds. Living near the ocean probably helped the birds survive, since the oceans didn't cool as much as areas farther inland. Still, the birds would have faced a real lack of food. How they survived is still an unsolved mystery.

Many insects survived the K/T extinction that followed the impact. Dragonflies (above) were one of the survivors. The dragonfly fossil below is from the early Cretaceous period, before the impact.

One of the biggest unknowns in the survival story is the season in which the meteorite hit. An impact in the spring would have shortened the growing season for plants in the Northern Hemisphere. Animals coming out of hiber-

Large fires happen today, though not of a size to match the K/T meteorite impact fires. Shown above is part of Yellowstone National Park during a devastating 1988 fire that consumed nearly one million acres.

nation would have faced a second winter. The Southern Hemisphere would have been less dramatically affected. Animals there would have been ready for winter. When they woke the next spring, temperatures would probably have been colder than usual, but at least if the skies had cleared, it would have been spring.

For animals coming out of hibernation, the world would have been a very different place for another reason. Nearly all the large predators would have disappeared. Freed from having to avoid hunters, little furry animals would have had a greater chance to multiply.

The blackened forests would soon have been full of the weeds that quickly sprout after fires. We know that ferns carpeted areas burned out by the impact's fireball, because large numbers of fern fossils have been found just above the impact debris layer.

The fossil record in the ocean shows a large crop, or "bloom," of microscopic plantlike creatures. After these plants were nearly wiped out after the impact, other creatures took their place in the changed waters, at least for a time. This blossoming went on for at least a couple of hundred thousand years—an amount of time much longer than humans have been on Earth. Slowly, little by little, new ecosystems—plants and animals living together in their environments—developed. Think of ecosystems as jigsaw puzzles of plants and animals living together. Many of the puzzle pieces vanished after the impact. It took many millions of years for new species to appear and make the puzzles whole again.

It's amazing how fast new life emerged. In the oceans, completely new species appeared within a few thousand years of the mass extinction, or even sooner—we can't be sure of the exact time. Freed from predators and faced with an unfamiliar world, new species of animals developed everywhere.

This great flowering of life after the extinction continued for millions of years. It took as long as ten million years for plant and animal life to return to pre-impact levels. On land the number of furry animal species increased rapidly in the two million years after the extinction, but nearly all the animals were small.

You may wonder if the mass extinction of dinosaurs really mattered. Suppose large dinosaurs were still around today. We'd probably have huge sauropods tromping through yards, munching on the shrubs of entire neighborhoods, and leaving bathtub-sized footprints in lawns and fields. And those are just the plant eaters.

People don't mix well with large predators. Today the only people who have to worry about being an animal's lunch are mostly Inuits fending off polar bears in the Arctic or divers swimming in shark-infested waters. It would have been tough for our ancestors if dinosaurs hadn't disappeared all those millions of years ago.

After the Yellowstone fire, changes in the types of plants that grew and their density caused different animals to come to the park, changing the area's ecosystem.

This is an artist's rendering of a multituberculate, an early type of mammal. During the Triassic to Cretaceous periods when it was common, it was prey to many creatures, including dinosaurs.

In fact, we may owe our very existence to that chunk of rock or ice that came crashing to Earth. Dinosaurs and the mouselike animals that were our very distant ancestors appeared on Earth at nearly the same time, some 230 million years ago. Both these small mammals and dinosaurs adapted to a great variety of ways of life. Yet it was the dinosaurs that became giants, while the species that ultimately led to us lived in the shadows, scurrying through tunnels and scrambling under the feet of the dinosaurs. This went on for 165 million years.

Near the end, dinosaurs were showing no signs of giving up their rule over the landscape. It's true that from the fossils we've found, it seems that the number of different kinds of dinosaurs decreased over several hundred thousand years before the impact. However, this may be only the way it looks to us. We simply haven't found enough dinosaur fossils yet to be sure. Even if the total number of dinosaurs was decreasing, dinosaurs had survived other hard times and had managed to flourish again and again.

How could small, furry mammals ever rise to greatness if they became a tasty snack whenever they came out of their hiding places? It's hard to topple a king who holds all the power. A slicing claw or a serrated tooth probably worked well to keep dinosaurs at the top of the food chain.

It is possible that the Ice Ages, periods of time after the K/T extinction when huge glaciers covered northern North America and northern Europe, would have ended the dinosaurs' reign anyway. Yet the dinosaurs had shown great adaptability to the cold. They had survived in the cold, dark reaches of Antarctica and along the shores of the Arctic seas. These were supremely capable creatures that might have developed an intelligence similar to our own,

given enough time. Without the chance collision of a rock from the asteroid belt or a "snowball" from the outer reaches of our solar system, evolution might have taken a very different path. We humans might not be here at all.

Changes in the environment caused by the K/T impact could have made it easier for small, furry creatures like the ancestors of this modern-day hedgehog to thrive.

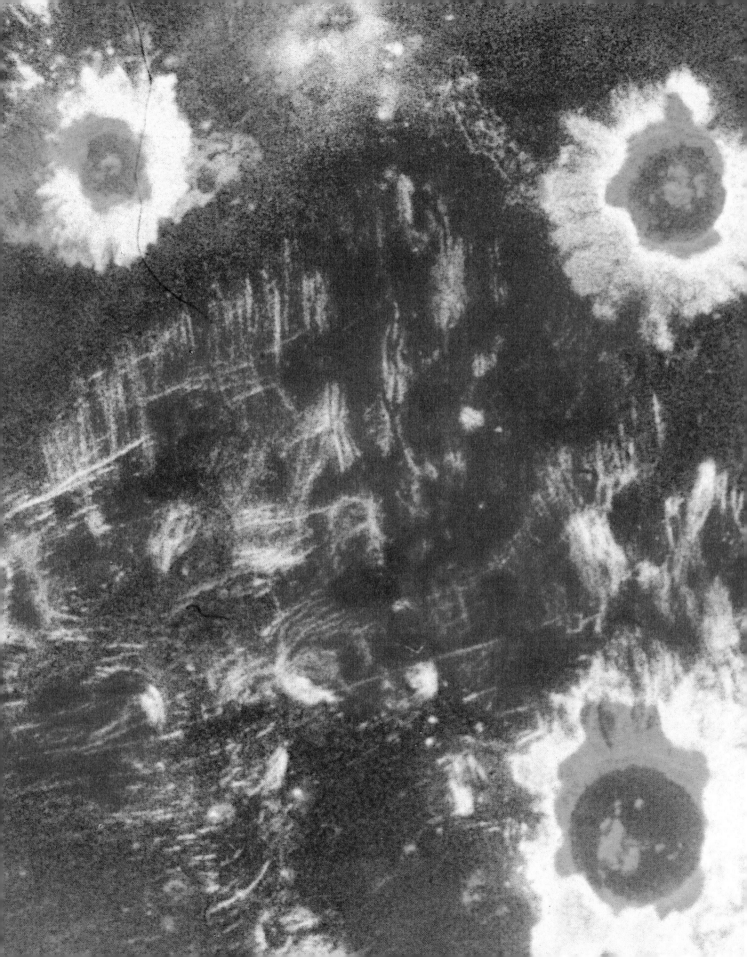

8

THE LONG RAIN OF ICE AND STONE

"It rained iron to the east of Lake Erh-hai.... Most of the people and animals struck by [the irons] were killed."—Yunnan Province, China, records from A.D. 1341

(above)
Meteorites have left large, easily visible impact craters on the moon.

(left)
Meteorites hit other planets as well. You can see several meteorite impact craters in this photograph of the surface of Venus.

What about a meteorite impact today? Could it happen? After all, the K/T meteorite wasn't even the biggest strike Earth has seen. In the dawn of our solar system, meteorite after meteorite pounded Earth. Though destructive, those large meteorite hits helped make life possible on our planet. A planet-sized asteroid strike helped form our moon, which is now responsible for the tides and for helping stir our oceans and the atmosphere. The continual crash of comets may also have resupplied Earth with water and other compounds that were blown away by large meteorite impacts while the oceans were still forming. But large impact events stopped more than three billion years ago.

Today, about ten tons of space rock—most the size of dust and sand—hit Earth every day, while more than 2,000 meteorites weighing more than about half a kilogram (a pound or so) hit the ground every year. How likely is it that one of these rocks could hit someone? Given the size of an average person and the number of people on Earth, calculations suggest that someone might be hit, but not necessarily hurt, approximately every 9 to 14 years.

❶ Clearwater Lakes, Canada
These craters were made by twin impacts 290 million years ago, something rarely found on Earth.

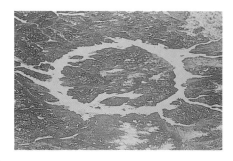

❷ Manicouagan, Canada
One of the largest impact craters still preserved on Earth's surface, this ice-covered crater is about 210 million years old and 70 kilometers (about 40 miles) in diameter.

❸ Lornar, India
Nearly 150 meters (about 500 feet) deep, the rim of this 50,000-year-old crater reaches about 20 meters (about 65 feet) high.

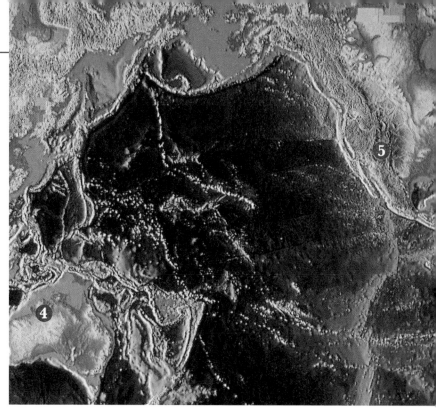

Chinese scholars and officials have been recording meteorite events since 645 B.C. Records from Yunnan Province in the year A.D. 1341 state, "It rained iron to the east of Lake Erh-hai. Houses and hilltops were damaged with holes. Most of the people and animals struck by them were killed." Fortunately there are no modern records of events like this, though there have been large meteor showers. These include a rain of more than 180,000 stones near Warsaw, Poland, in January 1868, and a shower of more than 14,000 fragments from a meteoroid breaking up in the sky over Holbrook, Arizona, in 1912.

Most of the asteroids whose orbits cross Earth's are small. Only a handful larger than half a kilometer (a quarter of a mile) across have orbits that intersect Earth's orbit. Should one of these asteroids hit us, it would do a lot of damage. Something like the meteorite hit that leveled part of Siberia is estimated to happen every two hundred to three hundred years. The world is more crowded now, and a Siberia-sized meteorite would do much more damage today than it did in 1908. Still, we

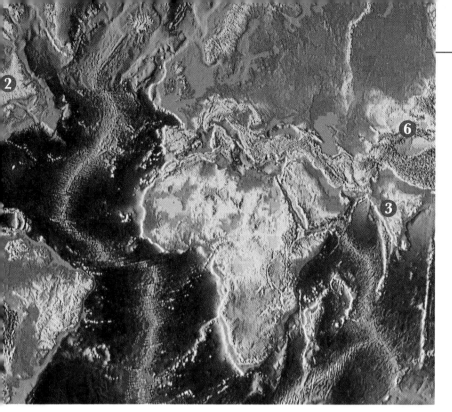

❹ Wolf Creek, Australia
The Wolf Creek crater is 300,000 years old and about 50 meters (about 165 feet) deep. It's partially buried under windblown sand.

❺ Arizona, USA
At almost 1.2 kilometers across, you could fit about 116 school buses end to end in this 25,000-year-old crater.

❻ Kara-Kul, Tadzhikistan
About 10 million years old and 45 kilometers (about 28 miles) in diameter, this crater is partly filled by a lake.

have yet to detect any asteroids or comets that could hit our planet and cause a K/T-sized extinction.

Comets and asteroids have given to our planet at the same time they have taken from it. Their long bombardment made the moon, delivered water from the heavens, and may have supplied Earth with materials that helped life begin. Some people have even suggested that rocks blasted off the surface of other planets may have delivered life here.

The rain of ice and stone has also destroyed life, perhaps many times. Early life forms may have been destroyed by giant impacts that boiled away the oceans and atmosphere. There is also good evidence that the course of life on Earth may have been changed by rocks from the skies. Large impact events about 35 million and 200 million years ago may be connected with extinctions, though not as dramatic as the end of the dinosaurs.

Giant impact craters are all around us despite the best efforts of time, water, and wind to erase them. Perhaps some of these craters also changed the course of ancient history. We are just beginning our search to find out.

GLOSSARY

asteroid A large rock in space from tens of meters up to tens of kilometers in diameter.

Chicxulub crater [chick-sue-LUBE] A buried impact crater in Mexico believed to be the place where a comet or asteroid hit 65 million years ago.

comet A collection of frozen gases, water, and rock from tens of meters up to tens of kilometers in diameter. Many comets have orbits very far from the sun, and they develop spectacular "tails" of vapor as they approach the inner solar system.

crater A hole blasted in the earth by an impact with an asteroid or comet.

Cretaceous period [kri-TAY-shuhs] The period of time between 144 million and 65 million years ago, when the last large dinosaurs lived on land.

drill string Long pieces of hollow pipe that are screwed together to create a continuous tube between the surface of the ocean and the seafloor.

ejecta [i-JEK-ta] Debris hurled from an impact crater.

extinction The death of an entire species.

foraminifera [fo-ram-uh-NIF-ahr-uh] Microscopic, single-celled creatures whose variety gives scientists clues about the history of life in the oceans.

fossil The remains, such as the skeleton or footprints, of ancient organisms.

impact A collision, such as the collision of rocks from space with the earth.

iridium A metal, similar to platinum, that is common in asteroids and meteoroids but is rare in rocks on Earth.

meteor [MEE-tee-or] The bright streak of light caused by a space rock burning up during entry into a planet's atmosphere.

meteorite [MEE-tee-uh-rite] A rock from space found on Earth's surface.

meteoroid [MEE-tee-uh-royd] A rock moving in space. Most range from dust-sized to sand-sized.

mutation An alteration, or change, in the structure of the genetic code of an organism by, for instance, exposure to sunlight or certain chemicals.

ozone An oxygen compound that absorbs harmful light from the sun.

Paleogene period [PAY-lee-oh-jeen] The period of time between 65 million to 23.7 million years ago, after the extinction of the dinosaurs.

paleontologist [pay-lee-ann-TAHL-uh-jist] A scientist who studies ancient nonhuman life.

radiation Energy in the form of light, x-rays, or radio waves. Some types of radiation are harmful to life.

sediment Sand, silt, or mud that has not been cemented to form solid rock.

shocked quartz Quartz crystals altered by tremendous pressure, such as that created by an impact event or nuclear explosion.

species A group of related organisms genetically similar enough to interbreed.

sulfuric acid A powerful acid. Sulfuric acid was formed when the K/T impact vaporized rocks in the Yucatán Peninsula.

tektite [TEK-tite] A glassy droplet formed when rock melts within a crater during an impact event.

tsunami [tsoo-NAH-me] A very destructive ocean wave caused by an impact or earthquake.

vents Cracks in the seafloor that let chemicals and minerals come up into the ocean.

Yucatán [you-cah-TAHN] A peninsula in southeastern Mexico that was the site of the K/T impact.

FURTHER READING

Barnes-Svarney, Patricia. *Asteroid: Earth Destroyer or New Frontier?*
New York: Plenum Press, 1996.

Benton, Michael. *How Do We Know Dinosaurs Existed?* Austin, TX:
Raintree/Steck-Vaughn, 1995.

Fagan, Elizabeth G. *Children's Atlas of Earth Through Time.* Chicago: Rand
McNally, 1990.

Kraske, Robert. *Asteroids: Invaders from Space.* New York: Aladdin Paper-
backs, 1998.

Marsh, Carole, and Arthur R. Upgren. *Asteroids, Comets, and Meteors (Secrets
of Space).* New York: Twenty-First Century Books, 1996.

Patent, Dorothy Hinshaw. *Fire: Friend or Foe?* New York: Clarion, 1998.

Pinet, Michele, and Alain Korkos. *Be Your Own Rock and Mineral Expert.*
New York: Sterling Publications, 1997.

Van Rose, Susanna, and Richard Bronson (illustrator). *Earth Atlas.*
New York: Dorling Kindersley, 1995.

INDEX

Acknowledgments

With particular appreciation to Dick Kroon and Adam Klaus, my co-conspirators on Ocean Drilling Program leg 171B, to the Ocean Drilling Program for making a lifetime of research possible, and to the National Science Foundation for its continued support for my research and that of many others on the curious and wonderful world we live in.

Credits

All photographs courtesy of Richard Norris/Woods Hole Oceanographic Institution, except for the following:

Cancalos, John/British Broadcasting Corporation Natural History Unit: 51 bottom; Clark, Chip/Smithsonian National Museum of History: 28 top; Corbis/Reuters: 45; Corbis/Royal Ontario Museum: 44; Department of Geology, National Museum of Wales: 2 middle, 30; Fogden, Michael and Patricia: 11 right; Gerlach, John/Animals, Animals: 51 top; Hamilton, Calvin J.: 57; Heinen, Guy, Luxembourg, Europe: 16, 17; Hillestad, Trond Eric: 4; Hurley, Jim/Krell Labs: 34 bottom, 35; Joint Oceanographic Institutions, Washington D.C.: 20; Kent, Breck P./Animals, Animals: 55; Kleindinst, Tom: 28 bottom; Lipschutz, Michael: 5 top; Martinez-Ruiz, Paqui/Universidad de Granada: 23 bottom right, 24 top and bottom right, 26, 27 inset; Mazzatenta, O. Louis/National Geographic Society: 48; McDonald, Kim/Chronicle of Higher Education: 9; NASA: front cover background, 1 background, 2 upper right, 8, 46, 56, 58–59; National Park Service/United States Department of the Interior: 13, 52, 53; Ocean Drilling Program: 21, 22, 23 top, bottom left, 24 left, 25, 27, 32–33; Oxford Scientific Films/Animals, Animals: 47; Perry, Marcus L., Spacewatch Project/Lunar and Planetary Laboratory, University of Arizona: 39; Royal Tyrell Museum, Alberta, Canada: front cover inset, 1 inset, 10; Rue, Leonard Lee III/Animals, Animals: 49; Satterwhite, Cecilia/NASA JSC: 5 bottom; Sharp, Tom/National Museum of Wales: 31; Schultz, Peter: 36, 37; Steinbrugge Collection, Earthquake Engineering Research Center/University of California, Berkeley: 43; Woods Hole Oceanographic Institution: back cover, 3 upper right, 29, 32–33; Winter, Vic and Jennifer Dudley/ICSTARS Astronomy: 2 upper left, 34 top left.

Illustrations on pages 3, 6, 12, 14, 15, 22, 26, 38, 40, 41, 42, 50, and 54 are by Greg Wenzel.